HOMESTEADING BOOK FOR BEGINNERS

Your Comprehensive Guide to Simple and Sustainable
Living for Beginner Homesteaders

LIVINUS O. MAGNUS

TABLE OF CONTENT

INTRODUCTION

In the quiet corners of quaint communities, where the gentle hum of nature intertwines with the aspirations of those seeking a simpler, more self-sufficient life, the allure of homesteading takes root. Picture a sun-kissed quarter acre, a canvas waiting for the strokes of your homesteading journey—a journey painted with the vibrant hues of fresh harvests, the comforting fragrance of natural soaps, and the rhythmic clucking of backyard chickens.

Now, imagine yourself standing at the threshold of this harmonious existence, eager yet perhaps a bit uncertain about where to begin. That's precisely where I found myself not too long ago, yearning for guidance to turn this dream into my reality.

In my quest, I stumbled upon a literary treasure, a beacon illuminating the path toward a fulfilling and sustainable life—the "Homesteading Book for Beginners." Its pages unfolded like a cherished map, revealing not just the destinations but the landscapes of knowledge, the hills and valleys of wisdom that lay ahead.

As I delved into the chapters, I was transported to the heart of homesteading, where the soil whispered secrets of successful cultivation, and the fragrance of essential oils lingered in the air, promising a natural and wholesome lifestyle. The journey, guided by the book's comprehensive wisdom, felt like a conversation with a trusted friend—a friend who understood the challenges of starting from scratch and the joy of seeing the first sprout peek through the soil.

From the simplicity of a single seed to the complexities of planning a thriving homestead, each turn of the page offered practical advice, relatable anecdotes, and vivid descriptions that resonated with the novice homesteader in me. Whether it was crafting herbal medicine or embracing micro-homesteading, the book unfolded like a narrative, weaving together the threads of information into a tapestry of empowerment and self-sufficiency.

So, fellow dreamer, if you stand at the edge of your own homesteading adventure, let "Homesteading Book for Beginners" be your compass, your guide through the uncharted territories of sustainable living. Join me as we turn the key to the homestead gate, unlocking the potential of our quarter acre, one chapter at a time. The journey awaits, and the possibilities are as vast as the homestead sky.

CHAPTER 1:
HOMESTEADING ESSENTIALS

Homesteading is a lifestyle that has evolved over centuries, rooted in self-sufficiency and a connection to the land. It involves individuals or families taking control of their living situation by producing their own food, generating energy, and, in some cases, even building their homes. In this chapter, we will explore the fundamental aspects of homesteading, delving into what it is, why people choose to homestead, and how this age-old practice has adapted to the demands of the modern world.

WHAT IS HOMESTEADING?

Homesteading is a lifestyle that centers around self-sufficiency, sustainability, and a strong connection to the land. At its core, homesteading involves individuals or families taking deliberate steps to reduce their reliance on external systems by producing their own food, generating energy, and creating a more self-sustaining living environment. This can extend to various aspects of life, including agriculture, animal husbandry, energy production, and crafting.

One of the essential elements of homesteading is the cultivation of a piece of land to meet the basic needs of the homesteaders. This often includes growing fruits, vegetables, and herbs, as well as raising livestock for meat, dairy, and other products. The goal is to create a closed-loop system where the homestead can produce most, if not all, of the necessities for daily living.

Homesteading is not a one-size-fits-all concept; it can take on different forms depending on the goals and preferences of the individuals involved. Some homesteads may focus on permaculture principles, emphasizing natural and sustainable farming practices. Others may prioritize off-grid living, utilizing alternative energy sources and reducing dependence on public utilities.

In addition to food production, many homesteaders engage in various DIY (do-it-yourself) projects, such as building their homes, creating homemade goods, and even generating their electricity. This hands-on approach fosters a sense of empowerment and independence, as homesteaders become intimately involved in the creation of their living environment.

Homesteading is also a mindset that values simplicity and sustainability. It encourages individuals to live in harmony with nature, respecting the ecosystem and minimizing their ecological footprint. By embracing a more intentional and mindful way of life, homesteaders often find a deeper connection to the environment and a sense of fulfillment in their daily activities.

WHY DO PEOPLE HOMESTEAD?
The decision to embrace homesteading is influenced by a combination of factors that vary from individual to individual. While there is no one-size-fits-all answer, several common motivations drive people to pursue a homesteading lifestyle.

Self-Sufficiency and Independence:

Homesteading offers a path to greater self-sufficiency and independence. By producing one's food, energy, and resources, individuals can reduce their reliance on external systems and gain a greater sense of control over their lives. This autonomy is particularly appealing to those who seek a more hands-on and self-directed way of living.

Connection to Nature:

Many homesteaders are drawn to the idea of reconnecting with nature. The modern urban lifestyle often alienates individuals from the natural world, and homesteading provides an opportunity to immerse oneself in the rhythms of the seasons, the earth, and the environment. This connection to nature is not only fulfilling but also essential for sustainable living practices.

Environmental Stewardship:

Homesteading aligns with principles of environmental stewardship. By adopting sustainable and regenerative practices, homesteaders actively contribute to the health of the planet. This includes organic farming, water conservation, and the reduction of waste. Homesteading is seen as a way to live in harmony with the Earth, minimizing the ecological impact of daily life.

Financial Freedom:

The cost of living continues to rise, prompting some individuals to explore homesteading as a means of achieving financial freedom. Producing one's food and generating energy on-site can significantly reduce monthly expenses. Moreover, the potential for selling surplus produce or handmade goods can create additional income streams for homesteaders.

Resilience in Uncertain Times:

Homesteading provides a level of resilience in the face of uncertainties such as economic downturns, natural disasters, or global crises. The ability to produce essential resources locally enhances a homesteader's capacity to weather challenges that may disrupt more centralized systems.

A Return to Simplicity:

In a world characterized by fast-paced living and constant connectivity, some individuals seek a simpler and more intentional lifestyle. Homesteading allows people to step away from the complexity of modern life, focusing on the basics and finding joy in the simplicity of daily tasks.

Health and Well-being:

Growing one's food often leads to a healthier and more sustainable diet. Homesteaders could cultivate organic produce, raise animals humanely, and make informed choices about the products they consume. The physical labor involved in homesteading also contributes to overall health and well-being.

While these motivations provide insight into why individuals choose to homestead, it's essential to recognize that each homesteader's journey is unique. The combination of these factors, along with personal values and goals, shapes the diverse tapestry of the homesteading community.

HOMESTEADING IN THE MODERN WORLD

Amid rapid technological advancements and urbanization, the practice of homesteading has not only persisted but has also adapted to meet the challenges and opportunities of the modern world. Modern homesteading incorporates traditional principles with contemporary innovations, allowing individuals to enjoy the benefits of self-sufficiency without completely disconnecting from the conveniences of the 21st century.

Technology and Innovation:

Modern homesteaders leverage technology to enhance their self-sufficiency efforts. From solar panels and wind turbines for off-grid energy to hydroponic systems for indoor gardening, technology plays a crucial role in optimizing resource utilization. Smart homesteading tools and applications also assist in monitoring and managing various aspects of a homestead efficiently.

Urban Homesteading:

Recognizing the challenges of space and resources in urban environments, a growing number of people are practicing urban homesteading. This involves integrating homesteading principles into city living, such as rooftop gardens, balcony farming, and small-scale animal husbandry. Urban homesteaders often focus on maximizing space and efficiency while maintaining a connection to sustainable living.

Community Building:

Homesteading is no longer confined to isolated rural settings. Modern homesteaders actively engage in community building, forming networks of like-minded individuals to share knowledge, resources, and support. Community-supported agriculture (CSA) models, communal gardening spaces, and online forums enable homesteaders to connect and collaborate, fostering a sense of shared purpose.

Entrepreneurship and Market Access:

The rise of e-commerce and farmers' markets provides modern homesteaders with avenues to turn their surplus produce and handmade goods into viable income streams. With increased awareness and demand for locally sourced, organic products, homesteaders can participate in the broader economy while maintaining their commitment to sustainable practices.

Education and Advocacy:

The digital age has facilitated the dissemination of information, allowing homesteaders to access a wealth of knowledge and share their experiences with a global audience. Online platforms, blogs, and social media enable homesteaders to educate and advocate for sustainable living practices, inspiring others to embrace aspects of the homesteading lifestyle.

Government and Policy Support:

In some regions, there is a growing recognition of the importance of sustainable and self-sufficient practices. Governments and local authorities are implementing policies to support small-scale agriculture, renewable energy initiatives, and environmentally conscious practices. This support helps create an enabling environment for modern homesteaders to thrive.

Environmental Conservation:

Modern homesteading aligns with contemporary concerns about environmental conservation. By adopting eco-friendly practices, such as rainwater harvesting, composting, and permaculture, homesteaders contribute to broader conservation efforts. This awareness of environmental impact reflects a commitment to leaving a positive legacy for future generations.

Homesteading in the modern world is a dynamic and adaptable lifestyle that combines traditional values with contemporary innovations. Whether practiced in rural or urban settings, homesteading continues to attract individuals seeking self-sufficiency, connection to nature, and a more intentional way of living. The fusion of age-old principles with modern tools and approaches ensures that homesteading remains a relevant and viable choice for those who value sustainability and independence in the 21st century.

CHAPTER 2:
CHOOSING YOUR HOMESTEAD: A BEGINNER'S GUIDE

Embarking on the journey of homesteading is a transformative experience that begins with a critical decision: choosing the right location for your homestead. This chapter serves as a comprehensive guide for beginners, exploring the essential factors to consider when selecting the perfect homestead location. From the geographical and climatic considerations to the legal and logistical aspects, this guide aims to provide aspiring homesteaders with the knowledge needed to make an informed decision and lay the foundation for a successful and fulfilling homesteading experience.

Exploring The Perfect Location for Your Homestead

Selecting the ideal location for your homestead is a multifaceted process that involves careful consideration of various factors. Your choice of location will significantly impact your ability to achieve self-sufficiency, cultivate a thriving environment, and integrate seamlessly into the homesteading lifestyle. In this section, we will delve into the key aspects to explore when determining the perfect location for your homestead.

Geographical Considerations:

The geographical features of your chosen location play a pivotal role in the overall success of your homestead. Consider the following aspects:

a. Soil Quality:

Assess the soil composition to determine its fertility and suitability for cultivation. Conduct soil tests to identify nutrient levels and potential challenges. Healthy, nutrient-rich soil is essential for successful gardening and farming.

b. Topography:

Examine the topography of the land, considering factors such as slope, elevation, and water drainage. While a slight slope can aid in water runoff, excessively steep terrain may pose challenges for construction and cultivation.

c. Water Availability:

Adequate and reliable water sources are crucial for a homestead. Assess the availability of water on the property, including wells, springs, rivers, or other water bodies. A sustainable and secure water supply is vital for irrigation, livestock, and household use.

d. Climate:

Understand the local climate patterns, including temperature ranges, precipitation, and seasonal variations. Choose a climate that aligns with your preferences and the types of crops and livestock you intend to raise. Consider the length of the growing season and the potential for extreme weather events.

Legal and Regulatory Considerations:

Navigating the legal and regulatory landscape is essential to ensure that your homesteading activities comply with local laws. Be mindful of the following:

a. Zoning Regulations:

Research local zoning regulations to understand how the land can be used. Some areas may have restrictions on agricultural activities, the size of structures, or the presence of certain animals. Ensure that your homesteading plans align with zoning requirements.

b. Building Codes:

Familiarize yourself with local building codes and permit requirements. Ensure that any structures you plan to build, whether a house, barn, or other facilities, adhere to safety standards and legal specifications. Obtain the necessary permits before starting construction.

c. Easements and Restrictions:

Investigate any easements or restrictions on the property. Easements grant others the right to use a portion of your land, while restrictions may limit certain activities. Understand the implications of any existing agreements on your homesteading plans.

d. Water Rights:

Clarify water rights associated with the property. In some regions, water rights may be a separate legal consideration. Ensure that you have legal access to the water sources on your land.

Logistical Considerations:

Practical and logistical considerations are essential for the day-to-day operations of your homestead. Pay attention to the following factors:

a. Access and Transportation:

Evaluate the accessibility of the property. Consider the quality of roads, proximity to essential services, and ease of transportation. Accessibility is crucial for receiving supplies, commuting, and marketing your homestead products.

b. Proximity to Markets:

Assess the proximity of your homestead to markets, both for selling your products and purchasing supplies. Consider transportation costs and the demand for locally produced goods in nearby communities.

c. Services and Amenities:

Determine the availability of essential services such as electricity, internet, and waste disposal. While homesteading often involves a degree of self-sufficiency, access to modern amenities can enhance comfort and convenience.

d. Wildlife and Pest Considerations:

Identify potential wildlife and pest challenges on the property. Certain regions may be more prone to specific pests or wildlife interactions that can impact crops and livestock. Develop strategies for pest control and protecting your homestead from wildlife.

Cultural and Community Considerations:

The cultural and community aspects of your chosen location can greatly influence your homesteading experience. Consider the following:

a. Community Support:

Explore the local community and assess its support for homesteading activities. A supportive community can provide valuable resources, shared knowledge, and a sense of belonging. Attend local events or join community groups to connect with like-minded individuals.

b. Cultural Fit:

Consider the cultural and social aspects of the area. Evaluate whether the local culture aligns with your values and preferences. Building connections with neighbors and becoming an active member of the community can enhance your overall homesteading experience.

c. Education and Healthcare:

Ensure that the location provides access to quality education and healthcare services if you have a family or plan to do so in the future. Consider the proximity of schools, hospitals, and other essential services.

d. *Local Economy:*

Examine the local economy and job market. While homesteading aims for self-sufficiency, having access to employment opportunities or local markets can be beneficial. Understanding the economic dynamics of the area provides insights into its long-term sustainability.

Financial Considerations:

The financial aspects of selecting a homestead location involve both upfront costs and long-term considerations. Pay attention to the following:

a. Land Cost and Financing:

Assess the cost of the land and explore financing options. Consider your budget and how it aligns with the available properties. Factor in additional costs such as property taxes and any potential development expenses.

b. Return on Investment:

Evaluate the potential return on investment for your homesteading activities. Consider the market demand for locally produced goods, the feasibility of selling surplus products, and the overall economic viability of your chosen location.

c. Resale Value:

While homesteading often involves a long-term commitment, it's wise to consider the resale value of the property. Factors such as location, accessibility, and the overall real estate market can influence the property's future value.

d. Cost of Living:

Examine the overall cost of living in the chosen area. Consider factors such as property taxes, utility costs, and the prices of goods and services. A lower cost of living can contribute to financial sustainability.

Choosing the perfect location for your homestead is a crucial step in realizing your homesteading dreams. The considerations outlined in this guide provide a comprehensive framework for assessing potential homestead locations, ensuring that you make an informed decision aligned with your goals and values. By thoroughly exploring the geographical, legal, logistical, cultural, and financial aspects, you can lay the groundwork for a successful and fulfilling homesteading journey.

CHAPTER 3:
HOMESTEAD PLANNING 101: SETTING UP FOR SUCCESS

Homestead planning is a foundational step on the path to a thriving and sustainable homestead. This chapter serves as a comprehensive guide, offering step-by-step strategies for organizing your homestead. From designing your layout to prioritizing essential infrastructure, this guide aims to equip homesteaders, both novice and experienced, with the knowledge needed to establish a well-organized and efficient homestead.

STEP-BY-STEP STRATEGIES FOR ORGANIZING YOUR HOMESTEAD

Homestead planning involves careful consideration of various elements to create a functional and harmonious living environment. This section breaks down the key aspects of homestead organization into step-by-step strategies, providing a roadmap for setting up a successful homestead.

Designing Your Homestead Layout:

Designing an effective layout is the first step in homestead planning. Consider the following strategies for optimizing your homestead design:

a. Zone Planning:

Utilize permaculture principles to create zones within your homestead. Zone 1 includes high-traffic and frequently used areas like the kitchen garden, while Zone 5 is left mostly wild. This zoning approach ensures efficient use of space and resources.

b. Access and Pathways:

Plan well-defined pathways and access points to facilitate movement around the homestead. Ensure that pathways are easily navigable, especially during adverse weather conditions. This design consideration enhances accessibility for daily chores and maintenance.

c. Sun and Wind Exposure:

Consider the orientation of your homestead concerning sunlight and wind exposure. Optimize the positioning of structures and planting areas to make the most of natural sunlight for gardening and to minimize exposure to harsh winds.

d. Water Management:

Plan for efficient water management by incorporating swales, rain gardens, and other water-harvesting techniques. Consider the natural flow of water on your property and design features that help capture and distribute water effectively.

Prioritizing Essential Infrastructure:

Establishing essential infrastructure is crucial for the smooth functioning of your homestead. Prioritize the following aspects during the planning phase:

a. Water Systems:

Develop a reliable water supply system. Install rainwater harvesting systems, wells, or other water sources based on the availability of water on your property. Ensure that water distribution systems reach key areas such as gardens, livestock areas, and the homestead itself.

b. Energy Sources:

Evaluate and choose appropriate energy sources for your homestead. This may include grid power, solar panels, wind turbines, or a combination of renewable energy sources. Design your energy infrastructure to meet both immediate and future needs.

c. Waste Management:

Implement effective waste management systems. This includes composting, recycling, and disposal methods for various types of waste generated on the homestead. Create designated areas for composting and recycling to minimize environmental impact.

d. Livestock Facilities:

If you plan to raise livestock, prioritize the design and construction of appropriate facilities. Consider barns, shelters, and secure enclosures for different types of animals. Ensure that these facilities promote the health and well-being of your livestock.

Establishing Garden Spaces:

Successful homesteading often involves cultivating a variety of crops to meet your dietary needs. Organize your garden spaces with the following strategies:

a. Crop Rotation Plans:

Implement crop rotation plans to optimize soil health and prevent the buildup of pests and diseases. Divide your garden into different plots for each season, rotating crops annually to maintain a balanced and nutrient-rich soil.

b. Companion Planting:

Use companion planting strategies to enhance the health and productivity of your crops. Planting complementary crops together can deter pests, improve soil fertility, and increase overall yield. Research and design your garden layout with companion planting principles in mind.

c. Raised Beds and Containers:

Consider incorporating raised beds and containers into your garden design. These structures provide better control over soil quality, drainage, and weed management. They are particularly useful in areas with challenging soil conditions.

d. Herb and Medicinal Gardens:

Allocate space for herb and medicinal gardens. Cultivate herbs for culinary use and medicinal plants for natural remedies. Integrating these gardens into your overall layout ensures easy access to these valuable resources.

Creating Functional Outdoor Spaces:

Homesteading is not only about productivity but also about creating enjoyable and functional outdoor living spaces. Consider the following strategies:

a. Outdoor Dining and Recreation Areas:

Designate outdoor spaces for dining and recreation. Create patios, decks, or designated areas for outdoor meals and relaxation. These spaces can serve as gathering points for family and friends.

b. Functional Landscaping:

Use landscaping to enhance both aesthetics and functionality. Plant fruit trees, shrubs, and ornamental plants strategically to provide shade, windbreaks, and visual appeal. Incorporate edible landscaping by integrating food-producing plants into decorative areas.

c. Fire Pit or Outdoor Cooking Area:

Include a fire pit or outdoor cooking area in your homestead layout. This provides a rustic and practical space for cooking, socializing, and enjoying the outdoors. Consider the placement to maximize safety and accessibility.

d. Storage Solutions:

Plan for adequate outdoor storage. This may include sheds, barns, or designated areas for storing tools, equipment, and supplies. Efficient storage solutions contribute to a clutter-free and organized homestead.

Infrastructure for Sustainable Living:

Homesteading often involves a commitment to sustainable living practices. Integrate the following strategies into your infrastructure planning:

a. Renewable Energy Systems:

Prioritize the use of renewable energy sources. Install solar panels, wind turbines, or other alternative energy systems to reduce reliance on conventional power sources. Consider energy-efficient appliances and lighting.

b. Composting Toilets and Greywater Systems:

Explore sustainable options for waste management. Composting toilets and greywater systems can reduce water consumption and contribute to nutrient cycling. Plan for the installation of these systems to minimize environmental impact.

c. Permaculture Principles:

Embrace permaculture principles in your homestead planning. Design your layout and systems to mimic natural ecosystems, promoting biodiversity, soil health, and resource efficiency. Implement permaculture techniques such as swales, guild planting, and integrated pest management.

d. Rainwater Harvesting:

Capture and utilize rainwater through efficient harvesting systems. Designate areas for rain barrels, cisterns, or other storage solutions to collect and store rainwater for use in gardens and other water-dependent areas.

Establishing a Functional Work Area:

Design a practical work area to facilitate daily homesteading tasks. Consider the following strategies:

a. Tool Storage and Organization:

Create designated spaces for tool storage and organization. Install racks, shelves, or tool sheds to keep essential equipment easily accessible and in good condition. A well-organized tool area enhances efficiency.

b. Workshop and Crafting Space:

Allocate space for a workshop or crafting area. This can be a separate structure or a designated section within an existing building. A functional workshop allows for DIY projects, repairs, and crafting endeavors.

c. Animal Care Facilities:

If you have livestock, plan for animal care facilities within easy reach of your work area. Designate spaces for feed storage, veterinary supplies, and grooming equipment.

Accessibility to these facilities streamlines daily care routines.

d. Emergency Preparedness:

Consider emergency preparedness in your homestead planning. Designate an area for storing emergency supplies, first aid kits, and communication tools. Establish evacuation plans and ensure that family members are familiar with safety protocols.

Budgeting and Timeline:

A well-defined budget and timeline are essential for successful homestead planning and implementation. Consider the following strategies:

a. Budget Allocation:

Break down your homestead project into individual components and allocate a budget to each. Include costs for land acquisition, infrastructure development, and ongoing maintenance. Be realistic about your financial capacity and prioritize essential elements.

b. Phased Implementation:

If necessary, consider phased implementation of your homestead plan. Prioritize critical components in the initial phases and allocate resources accordingly. This approach allows for gradual development and adjustments based on experience.

c. Contingency Planning:

Factor in contingency costs when creating your budget. Unforeseen challenges or changes in plans may arise, and having a contingency fund ensures that you can address unexpected expenses without compromising the overall project.

d. Timeline for Implementation:

Develop a realistic timeline for implementing your homestead plan. Consider the seasons, weather conditions, and the availability of resources. A well-structured timeline helps you stay organized and on track during the development process.

Homestead planning is a dynamic and integral aspect of successful homesteading. By following these step-by-step strategies, you can create a well-organized, efficient, and sustainable homestead that aligns with your goals and values. Whether you are a beginner or an experienced homesteader, careful planning lays the foundation for a fulfilling and thriving homesteading experience.

CHAPTER 4:
SUSTAINABLE LIVING DEMYSTIFIED

Sustainable living is more than a trend; it's a conscious choice to embrace practices that reduce environmental impact and promote a healthier, more balanced lifestyle. This chapter aims to demystify sustainable living, offering easy ways to incorporate eco-friendly practices and organic living into your daily routine. From understanding the principles of sustainability to practical tips for a greener home, this guide will empower you to make mindful choices that contribute to a more sustainable future.

EASY WAYS TO EMBRACE ECO-FRIENDLY PRACTICES AND ORGANIC LIVING
Understanding Sustainability:

Sustainability is at the core of eco-friendly living. It involves meeting the needs of the present without compromising the ability of future generations to meet their own needs. Explore the key components of sustainability:

a. Environmental Stewardship:

Embrace a sense of responsibility for the environment. Sustainable living involves making choices that minimize ecological impact, conserve natural resources, and promote biodiversity. Consider the lifecycle of products and the impact of your daily activities on the planet.

b. Social Equity:

Sustainable living extends beyond environmental concerns to social aspects. It includes promoting social equity, fair labor practices, and supporting communities. Make choices that contribute to a more just and equitable world, whether through ethical consumerism or community engagement.

c. Economic Viability:

Sustainable living also considers economic viability. Support businesses and practices that prioritize long-term economic health, fair wages, and ethical business practices. Consider the economic impact of your choices on local and global scales.

d. Personal Well-being:

Sustainable living recognizes the interconnectedness of personal well-being with that of the environment and society. Prioritize practices that enhance your physical and mental health, fostering a holistic approach to sustainable living.

Practical Tips for a Greener Home:

Transforming your home into a greener, more sustainable space involves simple yet impactful changes. Explore practical tips for an eco-friendlier home:

a. Energy Efficiency:

Invest in energy-efficient appliances, use LED light bulbs, and practice energy conservation habits. Simple measures like turning off lights and unplugging electronics when not in use contribute to reduced energy consumption.

b. Water Conservation:

Install water-efficient fixtures, fix leaks promptly, and practice water conservation in daily activities. Consider collecting rainwater for gardening and landscaping. Conserving water reduces strain on local water resources.

c. Waste Reduction:

Adopt a zero-waste mindset by reducing, reusing, and recycling. Choose products with minimal packaging, compost organic waste, and recycle materials whenever possible. Embrace a minimalist approach to reduce overall consumption.

d. Green Cleaning Products:

Transition to eco-friendly and non-toxic cleaning products. Many household cleaning items contain harmful chemicals that can negatively impact both the environment and your health. Opt for natural alternatives or make your own cleaning solutions.

Eco-Friendly Transportation Choices:

Transportation is a significant contributor to carbon emissions. Make conscious choices to reduce your carbon footprint:

a. Public Transportation:

Utilize public transportation whenever possible. Buses, trains, and other forms of mass transit are more energy-efficient and contribute to reduced traffic congestion.

b. Carpooling and Ridesharing:

Carpooling and ride-sharing are effective ways to decrease the number of vehicles on the road, leading to lower emissions. Coordinate with neighbors, coworkers, or friends for shared rides.

c. Biking and Walking:

Embrace active transportation by biking or walking for short distances. Not only does this reduce carbon emissions, but it also promotes personal health and well-being.

d. Electric and Hybrid Vehicles:

Consider electric or hybrid vehicles that have lower emissions compared to traditional gasoline-powered cars. As technology advances, electric vehicles become more accessible and affordable.

Mindful Consumer Choices:

Consumer choices have a direct impact on the environment. Adopt a mindful approach to consumption by considering the following:

a. Ethical and Sustainable Brands:

Support brands that prioritize ethical and sustainable practices. Look for certifications such as Fair Trade, Organic, or B Corp, indicating a commitment to social and environmental responsibility.

b. Local and Seasonal Eating:

Choose locally sourced and seasonal produce to reduce the carbon footprint associated with food transportation. Support local farmers' markets and community-supported agriculture (CSA) initiatives.

c. Clothing and Textiles:

Opt for sustainable and ethically produced clothing. Consider materials like organic cotton, hemp, or recycled fibers. Practice mindful wardrobe management by prioritizing quality over quantity.

d. Minimalism and Secondhand Shopping:

Embrace minimalism by decluttering and avoiding unnecessary purchases. Explore secondhand and thrift stores for clothing, furniture, and household items. Extend the lifecycle of products by choosing used items.

Gardening and Sustainable Agriculture:

Home gardening and sustainable agriculture contribute to both environmental and personal well-being. Explore eco-friendly practices for cultivating your own food:

a. *Organic Gardening:*

Practice organic gardening by avoiding synthetic pesticides and fertilizers. Embrace natural alternatives, companion planting, and crop rotation to maintain soil health.

b. Composting:

Start a composting system for kitchen scraps and yard waste. Compost enriches the soil, reduces the need for chemical fertilizers, and diverts organic waste from landfills.

c. Permaculture Principles:

Integrate permaculture principles into your garden design. Mimic natural ecosystems, promote biodiversity, and create sustainable food systems that work in harmony with the environment.

d. Rainwater Harvesting:

Install rain barrels or other rainwater harvesting systems to collect and store rainwater for irrigation. This reduces reliance on municipal water sources and conserves water resources.

Renewable Energy Solutions:

Transitioning to renewable energy sources is a key aspect of sustainable living. Explore the following options for incorporating renewable energy into your lifestyle:

a. Solar Panels:

Install solar panels on your property to harness the power of the sun for electricity. Solar energy is a clean and sustainable alternative to traditional grid power.

b. Wind Turbines:

In areas with sufficient wind, consider small-scale wind turbines for generating electricity. Wind energy is a viable option for off-grid living or supplementing traditional power sources.

c. Hydroelectric Power:

If you have access to a water source with sufficient flow, explore the possibility of small-scale hydroelectric power. This sustainable option harnesses the energy of flowing water.

d. Energy-Efficient Design:

Incorporate energy-efficient design principles into your home. Consider passive solar design, proper insulation, and strategic placement of windows to optimize natural light and reduce the need for artificial lighting and heating.

Educating and Advocating for Sustainability:

Spread awareness and advocate for sustainable living practices within your community. Education and advocacy play a crucial role in creating a widespread impact:

a. Community Workshops and Events:

Host or participate in community workshops and events focused on sustainability. Share knowledge about eco-friendly practices, showcase success stories, and engage in collaborative efforts.

b. Social Media and Online Platforms:

Utilize social media and online platforms to share information about sustainable living. Join or create groups dedicated to eco-friendly practices, where members can exchange ideas and support each other.

c. Supporting Environmental Initiatives:

Contribute to or support environmental initiatives and organizations. Participate in tree planting events, clean-up drives, or projects aimed at promoting sustainable practices in your local community.

d. Policy Advocacy:

Advocate for policies that promote sustainability at the local, national, and global levels. Stay informed about environmental policies and actively participate in discussions and advocacy efforts. Sustainable living is a journey that begins with small, mindful choices and can lead to a profound impact on the planet. By incorporating eco-friendly practices and organic living into your daily routine, you contribute to a more sustainable and resilient future. Whether through energy-efficient choices, mindful consumption, or advocating for change, each action plays a part in demystifying sustainable living and creating a positive ripple effect in the world.

CHAPTER 5:
DIY ON THE HOMESTEAD: PRACTICAL SOLUTIONS FOR BEGINNERS

Engaging in do-it-yourself (DIY) projects on the homestead is not just a cost-saving strategy; it's a pathway to self-sufficiency and a deeper connection with your land. This chapter delves into practical solutions for beginners, offering budget-friendly tips and tricks to empower homesteaders in their DIY endeavors. From building structures to crafting homemade essentials, this guide aims to inspire and equip individuals with the skills and knowledge needed to enhance their self-sufficient life on the homestead.

Budget-Friendly Tips and Tricks For A Self-Sufficient Life

DIY Structures and Shelters:

Building structures on the homestead, whether for housing, storage, or livestock, is a rewarding and cost-effective venture. Explore the following tips for DIY structures:

a. Basic Carpentry Skills:

Acquiring basic carpentry skills is fundamental for any DIY enthusiast. Invest time in learning how to measure, cut, and assemble wood effectively. Online tutorials, books, and community workshops are valuable resources for honing carpentry skills.

b. Simple Shelter Designs:

Start with simple shelter designs that match your needs. A chicken coop, tool shed, or a basic greenhouse are excellent beginner projects. These structures not only provide functional spaces but also allow you to practice essential construction skills.

c. Recycled and Salvaged Materials:

Embrace sustainability by using recycled and salvaged materials for your DIY projects. Old pallets, reclaimed wood, and salvaged windows can be repurposed into functional and aesthetically pleasing structures. This not only reduces costs but also minimizes environmental impact.

d. Community Collaboration:

Collaborate with the homesteading community. Exchange skills and resources with neighbors or fellow homesteaders. Joint projects can be not only cost-effective but also foster a sense of community and shared knowledge.

Homemade Tools and Equipment:

Crafting your own tools and equipment adds a personalized touch to your homestead while saving on costs. Consider the following DIY tips for crafting essentials:

a. Blacksmithing Basics:

Explore blacksmithing basics to create your own tools. A simple forge setup and anvil can enable you to craft items such as hooks, garden trowels, and even basic farm implements. Blacksmithing is a skill that evolves with practice.

b. Woodworking for Tools:

Woodworking is another versatile skill for crafting tools. Wooden handles for garden tools, wooden mallets, or even custom-sized trays and containers can be created with basic woodworking tools.

c. DIY Power Tools:

Consider making your own manual power tools. A foot-powered lathe or a pedal-operated grinding wheel are examples of DIY alternatives that can be effective for various tasks around the homestead.

d. Tool Repair and Maintenance:

Learn basic tool repair and maintenance. Regularly sharpening blades, fixing handles, and addressing minor issues can extend the lifespan of your tools, reducing the need for frequent replacements.

Home Improvement and Repairs:

Keeping your homestead in good repair is crucial for its long-term sustainability. Develop your home improvement skills with these DIY tips:

a. Basic Plumbing Repairs:

Familiarize yourself with basic plumbing repairs. Fixing leaks, replacing faucets, and understanding your septic system are essential skills that can save money on professional services.

b. Electrical Work for Beginners:

Start with simple electrical projects, such as replacing outlets or light switches. Ensure you follow safety protocols and consult with professionals when necessary. Understanding the basics of your electrical system is empowering.

c. Roofing and Gutter Maintenance:

Learn basic roofing skills for minor repairs and gutter maintenance. Keeping your roof in good condition prevents costly damage to the structure and interior spaces.

d. Weatherproofing and Insulation:

Weatherproofing and adding insulation to your home contribute to energy efficiency. Seal gaps, use weatherstripping, and insulate attics and walls to regulate indoor temperatures and reduce heating or cooling costs.

DIY Gardening and Farming Solutions:

Enhance your self-sufficiency in gardening and farming by incorporating DIY solutions. From irrigation to composting, these tips can make a substantial difference:

a. Homemade Irrigation Systems:

Craft homemade irrigation systems using readily available materials. Drip irrigation using recycled bottles, soaker hoses, or even rain barrel irrigation systems can efficiently water your crops without significant expense.

b. Composting Techniques:

Explore different composting techniques suited to your space and needs. Vermicomposting, trench composting, or DIY compost bins can turn kitchen and yard waste into nutrient-rich soil amendments.

c. DIY Garden Tools:

Create your own garden tools using basic materials. Wooden plant markers, seedling pots made from newspaper, or a simple dibber for planting are easy projects that contribute to a well-equipped garden.

d. Vertical Gardening Structures:

Maximize space by constructing vertical gardening structures. Vertical planters, trellises, and hanging gardens are DIY projects that allow you to grow more in limited space, especially useful for urban or small-scale homesteads.

DIY Renewable Energy Projects:

Embracing renewable energy sources on the homestead can reduce reliance on traditional power grids. Explore DIY projects in renewable energy:

a. Solar Water Heaters:

Build a solar water heater using simple materials. This DIY project can supplement your hot water needs, particularly in sunny climates. Online guides and community forums provide step-by-step instructions.

b. DIY Solar Dehydrator:

Create a solar dehydrator to preserve fruits, vegetables, and herbs. This sustainable solution harnesses the power of the sun for food preservation, reducing the need for electric dehydrators.

c. Wind-Powered Generator:

Consider building a small-scale wind-powered generator. While larger wind turbines may require more expertise, DIY projects for small-scale generators can provide supplementary power for specific needs.

d. DIY Rainwater Harvesting Systems:

Build your own rainwater harvesting system using simple materials. From gutter collection systems to storage barrels, DIY rainwater harvesting projects can supply water for gardens and livestock.

Homemade Household Products:

Crafting your own household products not only reduces reliance on store-bought items but also minimizes exposure to harmful chemicals. Try these DIY alternatives:

a. Natural Cleaning Solutions:

Create natural cleaning solutions using common household ingredients like vinegar, baking soda, and lemon. DIY cleaners are cost-effective, eco-friendly, and safer for your health.

b. Homemade Personal Care Products:

Experiment with making your own personal care products. DIY toothpaste, deodorant, and shampoo often involve simple recipes with natural ingredients, reducing the need for commercially produced items.

c. Beeswax Wraps and DIY Cloth Products:

Reduce plastic use by making your own beeswax wraps for food storage. Additionally, consider DIY cloth products like reusable cloth napkins, dishcloths, and cleaning wipes.

d. Homemade Candles and Air Fresheners:

Craft your own candles using natural waxes and essential oils. Homemade air fresheners using baking soda, essential oils, and dried herbs provide a pleasant and toxin-free alternative.

Skills Development and Continuous Learning:

Embracing a DIY lifestyle on the homestead involves ongoing skills development and continuous learning. Cultivate a mindset of curiosity and self-improvement with these tips:

a. Online Resources and Courses:

Take advantage of online resources and courses. Platforms like YouTube, Udemy, and Skillshare offer tutorials on a wide range of DIY topics, from carpentry to permaculture.

b. Community Workshops and Classes:

Participate in local community workshops and classes. Many communities offer hands-on sessions on various DIY skills. Check community centers, agricultural extension offices, or local homesteading groups for information on upcoming events.

c. Apprenticeships and Mentorships:

Seek apprenticeships or mentorships with experienced individuals in specific DIY fields. Learning directly from someone with expertise can accelerate your skill development and provide valuable insights.

d. Practice and Experimentation:

Practice is key to mastering any skill. Experiment with different DIY projects on a small scale before taking on larger endeavors. Learn from both successes and challenges to refine your techniques.

DIY on the homestead is not just a practical approach to budgeting; it's a lifestyle that fosters independence, creativity, and a deeper connection to the land. By incorporating these budget-friendly tips and tricks, beginners can embark on a fulfilling journey of self-sufficiency. Whether constructing structures, crafting tools, or developing renewable energy solutions, each DIY project contributes to a more resilient and sustainable homestead life.

CHAPTER 6:
FROM SEED TO TABLE: GROWING YOUR OWN GROCERIES

Embarking on the journey from seed to table is a rewarding endeavor that brings you closer to your food source, enhances sustainability, and fosters a deeper connection with the land. In this chapter, we explore beginner-friendly tips for successful farming, guiding you through the various stages of cultivating your own groceries. From selecting seeds to harvest and meal preparation, this comprehensive guide equips you with the knowledge needed to turn your homestead into a thriving source of fresh, homegrown produce.

Beginner-Friendly Tips for Successful Farming

Choosing the Right Seeds:

Selecting the right seeds is the first step towards a successful harvest. Consider the following tips for choosing seeds for your home garden:

a. Climate and Soil Considerations:

Choose seeds that are well-suited to your local climate and soil conditions. Different plants thrive in varying temperature ranges and soil types, so research which crops are best suited for your region.

b. Heirloom and Open-Pollinated Varieties:

Opt for heirloom and open-pollinated seed varieties. These seeds produce plants that closely resemble their parent plants, allowing you to save seeds from year to year. This promotes biodiversity and helps preserve traditional plant varieties.

c. Seed Quality and Viability:

Assess the quality and viability of seeds before purchasing. Look for reputable seed suppliers and check for information on germination rates. High-quality seeds significantly contribute to the success of your farming endeavors.

d. Start with Easy-to-Grow Crops:

If you're a beginner, start with easy-to-grow crops. Herbs like basil, vegetables like tomatoes and zucchini, and salad greens are generally forgiving and provide a good introduction to home gardening.

Planning Your Garden Layout:

Efficient planning of your garden layout maximizes space, sunlight, and accessibility. Follow these tips to design a well-organized and productive garden:

a. Companion Planting:

Implement companion planting strategies to enhance plant growth and deter pests. Certain plants thrive when planted together, creating mutually beneficial relationships. For example, planting basil near tomatoes can improve tomato flavor and discourage pests.

b. Crop Rotation:

Plan for crop rotation to prevent soil depletion and control pests and diseases. Avoid planting the same family of crops in the same location year after year. Rotate crops to maintain soil fertility and reduce the risk of soil-borne pathogens.

c. Raised Beds and Containers:

Consider using raised beds and containers, especially if you have limited space or poor soil quality. Raised beds provide better control over soil conditions, drainage, and weed management. Containers are suitable for growing herbs, small vegetables, and even fruit trees.

d. Vertical Gardening:

Utilize vertical gardening techniques to maximize space. Vertical structures like trellises, vertical planters, and hanging baskets allow you to grow more crops in limited space, making them ideal for small or urban homesteads.

Preparing the Soil:

Soil preparation is a critical factor in successful farming. Follow these guidelines to ensure your soil is nutrient-rich and conducive to plant growth:

a. Soil Testing:

Conduct a soil test to determine its pH and nutrient levels. Soil testing kits are readily available and provide insights into any amendments needed. Adjusting the pH and nutrient balance creates an optimal environment for plant roots.

b. Adding Organic Matter:

Enhance soil fertility by adding organic matter such as compost, well-rotted manure, or cover crops. Organic matter improves soil structure, water retention, and nutrient availability, fostering a healthy growing environment.

c. Mulching:

Apply mulch to retain moisture, suppress weeds, and regulate soil temperature. Organic mulches like straw, wood chips, or leaves also contribute to soil health as they break down over time, adding more organic matter.

d. No-Till Gardening:

Consider adopting no-till gardening practices. No-till methods preserve soil structure and minimize disturbance to beneficial soil organisms. This approach promotes long-term soil health and reduces the risk of erosion.

Planting and Transplanting:

Successful planting and transplanting set the stage for a bountiful harvest. Follow these tips to ensure your plants get a healthy start:

a. Follow Planting Guidelines:

Adhere to recommended planting guidelines for each crop. Pay attention to spacing, depth, and timing. This information is usually available on seed packets or plant labels.

b. Hardening Off Transplants:

If you're transplanting seedlings started indoors, gradually acclimate them to outdoor conditions through a process called hardening off. Expose them to increasing periods of sunlight and outdoor conditions over several days before planting them in the garden.

c. Companion Planting Strategies:

Implement companion planting strategies during the planting phase. Certain plants have natural affinities for each other, helping with nutrient uptake and pest control. For example, planting marigolds with tomatoes can deter nematodes.

d. Watering Wisely:

Water newly planted seeds and transplants carefully. Provide consistent moisture without overwatering. Use methods like drip irrigation or watering at the base of plants to minimize water on foliage, reducing the risk of diseases.

Caring for Your Garden:

Ongoing care is essential for a thriving garden. Follow these tips to ensure your plants remain healthy throughout the growing season:

a. Weeding and Mulching:

Regularly weed your garden to prevent competition for nutrients and space. Mulching helps suppress weeds and retains soil moisture, reducing the need for excessive watering.

b. Pruning and Thinning:

Practice pruning and thinning to optimize plant growth. Remove dead or diseased foliage, and thin overcrowded plants to allow for better air circulation. This reduces the risk of fungal diseases and promotes healthier plants.

c. Fertilizing:

Monitor plant nutrient needs and fertilize accordingly. Use organic fertilizers, compost, or natural amendments to provide a balanced nutrient profile. Avoid over-fertilizing, as this can lead to nutrient imbalances and negatively impact plant health.

d. Pest and Disease Management:

Implement integrated pest management strategies to control pests and diseases. Encourage natural predators, use companion planting, and employ organic pest control methods to minimize the need for chemical interventions.

Harvesting Your Bounty:

The joy of harvesting your own produce is a highlight of home gardening. Follow these guidelines for a successful and satisfying harvest:

a. Harvest at the Right Time:

Harvest crops at their peak ripeness for the best flavor and nutritional content. Different crops have specific indicators of readiness, such as color, size, or texture. Refer to gardening guides or seed packets for guidance.

b. Use Proper Harvesting Techniques:

Employ proper harvesting techniques to avoid damage to plants. Use sharp, clean tools for cutting stems or picking fruits to minimize stress on the plant and reduce the risk of introducing diseases.

c. Preserving and Storing:

Explore various methods of preserving and storing your harvest. Canning, freezing, dehydrating, and root cellaring are effective ways to extend the shelf life of fruits, vegetables, and herbs, allowing you to enjoy your homegrown produce year-round.

d. Seed Saving:

Consider saving seeds from open-pollinated and heirloom varieties. Properly collected and stored seeds can be used for future plantings, promoting seed sovereignty and preserving unique plant varieties.

Cooking and Meal Preparation:

Bringing your homegrown produce to the table is the final, delightful step in the process. Discover creative and satisfying ways to incorporate your harvest into meals:

a. Seasonal Cooking:

Embrace seasonal cooking by planning meals around the crops available in your garden. Seasonal eating not only maximizes flavor but also connects you with the natural rhythm of your homestead.

b. Preserving the Harvest:

Utilize preserved items in your cooking. Incorporate canned tomatoes into sauces, use dried herbs for seasoning, and enjoy frozen fruits in smoothies. Preserving allows you to enjoy the flavors of your garden throughout the year.

c. Experimenting with Homegrown Herbs:

Explore the diverse flavors of homegrown herbs. Experiment with fresh herbs to enhance the taste of your dishes. From basil and cilantro to rosemary and thyme, herbs add depth and complexity to meals.

d. Sharing the Bounty:

Share the abundance of your harvest with neighbors, friends, and community members. A sense of community builds around sharing homegrown produce, fostering connections and creating a network of support.

Growing your own groceries from seed to table is a fulfilling and empowering journey. By following these beginner-friendly tips for successful farming, you can cultivate a productive and sustainable garden, providing fresh, nutritious produce for yourself and your community. From the careful selection of seeds to the joy of cooking with your homegrown ingredients, each step contributes to a more self-sufficient and connected way of living on the homestead.

CHAPTER 7:
BUILDING YOUR HOMESTEAD COMMUNITY:
FINDING YOUR TRIBE

Embarking on a homesteading journey is not just about cultivating the land; it's also about fostering connections with like-minded individuals who share a passion for sustainable living. In this chapter, we explore the importance of building a homestead community and provide guidance on finding your tribe. From connecting with neighbors to engaging with online homesteading forums, this comprehensive guide aims to help you create a supportive network that enhances your homesteading experience.

Connecting with Like-Minded Individuals Locally and Online

Understanding the Importance of Community:

Building a homestead community goes beyond the practicalities of sharing resources; it's about creating a network that enriches your homesteading journey. Here's why community is vital:

a. Shared Knowledge and Experience:

A homestead community provides a wealth of shared knowledge and experience. Whether you're a seasoned homesteader or a beginner, learning from others' successes and challenges can significantly enhance your skills and decision-making.

b. Resource Sharing:

Community members often engage in resource sharing, from tools and equipment to seeds and surplus produce. This collaborative spirit fosters a sense of interconnectedness and reduces individual costs.

c. Emotional Support:

Homesteading can be both rewarding and demanding. Having a community to share triumphs, setbacks, and the inevitable learning experiences provides essential emotional support. It's a network of individuals who understand the unique challenges and joys of homesteading.

d. Community Resilience:

In times of unforeseen challenges, such as extreme weather events or other emergencies, a homestead community provides a safety net. Collective skills and resources contribute to community resilience, ensuring that members can weather difficulties more effectively together.

Connecting with Local Homesteaders:

Establishing connections with local homesteaders is a valuable aspect of building your homestead community. Follow these steps to connect with like-minded individuals in your area:

a. Attend Local Events:

Look for local events, workshops, or farmers' markets where homesteaders might gather. These gatherings provide opportunities to meet and connect with fellow homesteaders, share experiences, and learn from each other.

b. Join Community Organizations:

Explore community organizations related to agriculture, sustainability, or homesteading. Joining such groups introduces you to local individuals who share your interests. Check local community centers, gardening clubs, or environmental organizations.

c. Community Gardens:

Participate in community gardens if available in your area. These shared spaces not only provide a platform for cultivating plants but also offer a chance to interact with other gardeners and homesteaders.

d. Start or Join a Homesteading Meetup:

If there isn't already a homesteading meetup in your community, consider starting one. Websites like Meetup.com can help you connect with potential members. Alternatively, join existing local homesteading groups.

Engaging with Online Homesteading Communities:

The digital age has opened avenues for connecting with homesteaders worldwide. Online communities offer a virtual space to share knowledge, seek advice, and build connections:

a. Join Homesteading Forums:

Participate in homesteading forums and discussion boards. Platforms like Homesteading Today, The Homesteading Forum, or BackYardHerds bring together a diverse community of homesteaders sharing insights and experiences.

b. Social Media Groups:

Join homesteading groups on social media platforms like Facebook, Instagram, or Reddit. These groups are excellent for real-time interactions, sharing photos, and seeking advice. Look for groups that align with your specific interests, such as poultry keeping or permaculture.

c. Blogging and Online Communities:

Follow homesteading blogs and engage with the online community through comments and discussions. Many homesteaders share their experiences, challenges, and solutions through blogs, fostering a sense of virtual camaraderie.

d. Webinars and Virtual Events:

Attend webinars and virtual events organized by homesteading experts or organizations. These online gatherings provide opportunities to connect with like-minded individuals, ask questions, and learn from experienced homesteaders.

Starting a Homesteading Cooperative:

A homesteading cooperative is a collaborative effort where community members pool resources, skills, and efforts for mutual benefit. Here's how you can initiate or join a homesteading cooperative:

a. Identify Common Interests:

Find individuals in your community or online who share similar homesteading goals or interests. Common interests form the foundation for a successful cooperative.

b. Define Roles and Contributions:

Clearly define roles and contributions within the cooperative. Determine how resources, tasks, and responsibilities will be shared. This clarity minimizes misunderstandings and ensures equitable participation.

c. Establish Communication Channels:

Set up effective communication channels for the cooperative. Whether through regular meetings, online platforms, or a combination of both, clear communication is essential for successful collaboration.

d. Start Small and Expand:

Begin with small, manageable projects to test the effectiveness of the cooperative. As trust and cooperation grow, the cooperative can gradually expand its scope and take on more ambitious ventures.

Organizing Homesteading Workshops and Skill Shares:

Organizing workshops and skill shares within your community creates opportunities for learning and collaboration. Here's how you can initiate these events:

a. Identify Skills within the Community:

Assess the skills and expertise within your homestead community. Identify individuals who excel in specific areas, whether it's beekeeping, cheese-making, or carpentry.

b. Plan and Coordinate Events:

Plan and coordinate workshops or skill shares based on the identified expertise. This could include hands-on activities, demonstrations, or informational sessions. Rotate the responsibility for organizing events among community members.

c. *Utilize Community Spaces:*

Utilize community spaces such as community centers, local libraries, or outdoor areas for hosting workshops. Collaborate with local authorities or organizations to secure suitable venues.

d. Encourage Participation and Feedback:

Encourage active participation from community members and seek feedback to improve future events. The goal is to create an inclusive and supportive environment where everyone can contribute and learn.

Creating a Homesteading Co-Housing Community:

Homesteading co-housing involves living in proximity with other homesteaders, sharing resources, and creating a self-sufficient community. Here's a guide to creating a homesteading co-housing community:

a. Identify Potential Members:

Identify individuals who share a vision for homesteading co-housing. Look for those with complementary skills, interests, and values to create a diverse and resilient community.

b. Define Shared Goals and Values:

Clearly define shared goals and values for the co-housing community. Discuss expectations regarding resource sharing, responsibilities, and communal decision-making.

c. Secure Suitable Land:

Identify and secure suitable land for the co-housing community. Consider factors such as location, soil quality, and zoning regulations. Collaborate with a real estate professional experienced in sustainable land acquisition.

d. *Design Sustainable Structures:*

Work with architects and builders to design sustainable and efficient structures for the co-housing community. Incorporate eco-friendly features, communal spaces, and areas for shared activities.

Navigating Challenges and Conflict Resolution:

Building a homestead community is not without challenges. Conflict resolution is a crucial skill to maintain harmony. Consider the following strategies:

a. Open Communication:

Foster open and honest communication within the community. Encourage members to express concerns or conflicts early on to prevent issues from escalating.

b. Establish Clear Guidelines:

Develop clear guidelines and rules for the community. Clearly outline expectations regarding shared resources, responsibilities, and decision-making processes. This clarity minimizes misunderstandings.

c. Mediation and Facilitation:

In the event of conflicts, consider mediation or facilitation. A neutral third party can help guide discussions and find mutually agreeable solutions. This could be a community member trained in conflict resolution or an external mediator.

d. Learning from Challenges:

View challenges as opportunities for growth and learning. Reflect on conflicts to understand their root causes and use them as lessons to strengthen the community's resilience.

Building your homestead community is a transformative aspect of the homesteading journey. Whether connecting with local homesteaders or engaging in online communities, the shared knowledge, resources, and emotional support foster a sense of belonging and resilience. Initiating or participating in homesteading cooperatives, workshops, co-housing communities, and navigating challenges with effective communication all contribute to a robust and thriving homestead community. As you embark on this communal journey, remember that building connections with like-minded individuals is not just about shared resources; it's about cultivating a network that enriches and sustains the collective spirit of homesteading.

CHAPTER 8:
GREEN HOMESTEAD: ECO-FRIENDLY INFRASTRUCTURE BASICS

Creating a green homestead involves harmonizing with the environment, embracing sustainable practices, and prioritizing energy efficiency. This chapter delves into the fundamentals of eco-friendly infrastructure, offering simple steps for sustainable building and energy efficiency. From choosing green materials to implementing renewable energy solutions, this comprehensive guide empowers homesteaders to reduce their environmental impact and cultivate a more sustainable lifestyle.

Simple Steps for Sustainable Building and Energy Efficiency

Choosing Green Building Materials:

The materials used in constructing your homestead play a crucial role in its environmental footprint. Opting for green building materials contributes to sustainability and reduces the ecological impact. Consider the following options:

a. Recycled and Reclaimed Materials:

Utilize recycled and reclaimed materials for construction. Reclaimed wood, salvaged bricks, or recycled metal can be repurposed into building components, reducing the demand for new resources.

b. Bamboo:

Bamboo is a rapidly renewable resource with excellent strength and versatility. It can be used for various construction purposes, from flooring to structural elements, providing a sustainable alternative to traditional wood.

c. Sustainable Timber:

If using wood, choose sustainably sourced timber certified by organizations like the Forest Stewardship Council (FSC). This ensures that the wood comes from responsibly managed forests, promoting conservation and regeneration.

d. Natural Insulation Materials:

Opt for natural insulation materials like wool, cotton, or cellulose instead of synthetic options. These materials have a lower environmental impact, provide effective insulation, and are often more energy efficient.

Energy-Efficient Design Principles:

Designing your homestead with energy efficiency in mind is a cornerstone of a green lifestyle. Implementing energy-efficient principles not only reduces your environmental footprint but also lowers utility costs. Consider the following design strategies:

a. Passive Solar Design:

Embrace passive solar design principles to maximize natural heating and lighting. Orientate your buildings to capture sunlight, incorporate large south-facing windows, and use thermal mass materials to store and release heat gradually.

b. Proper Insulation:

Ensure proper insulation throughout your homestead. Well-insulated walls, roofs, and floors prevent heat loss in the winter and keep interiors cool in the summer, reducing the need for excessive heating or cooling.

c. Energy-Efficient Windows:

Invest in energy-efficient windows with low-emissivity coatings and double or triple glazing. These windows reduce heat transfer, provide better insulation, and contribute to overall energy savings.

d. Ventilation and Airflow:

Implement effective ventilation strategies to maintain good indoor air quality. Proper airflow reduces the reliance on mechanical cooling systems, promoting a healthier and more energy-efficient living environment.

Off-Grid and Renewable Energy Solutions:

Embracing off-grid and renewable energy solutions is a pivotal step towards a sustainable homestead. Generate your own energy from renewable sources to reduce dependence on conventional power grids. Consider the following options:

a. Solar Power:

Install solar panels to harness the power of the sun. Solar energy is a clean and renewable source that can provide electricity for lighting, appliances, and other energy needs on the homestead.

b. Wind Turbines:

If your location permits, consider installing a small-scale wind turbine. Wind energy can complement solar power, especially in areas with consistent wind patterns, providing additional renewable energy for your homestead.

c. Hydroelectric Systems:

For homesteads located near flowing water, small-scale hydroelectric systems can be a sustainable energy solution. These systems convert the kinetic energy of flowing water into electricity.

d. Backup Power Storage:

Implement energy storage solutions like batteries to store excess energy generated by renewable sources. This ensures a reliable power supply during periods of low energy production or grid outages.

Water Conservation Practices:

Efficient water use is integral to a green homestead. Implementing water conservation practices helps preserve this precious resource and reduces the environmental impact of water consumption. Consider the following strategies:

a. Rainwater Harvesting:

Install rainwater harvesting systems to collect and store rainwater for irrigation, livestock, and household use. This sustainable practice reduces reliance on traditional water sources and helps conserve local water supplies.

b. Drip Irrigation:

Opt for drip irrigation systems in gardens and crop areas. Drip systems deliver water directly to the base of plants, minimizing water wastage through evaporation or runoff compared to traditional overhead watering methods.

c. Greywater Recycling:

Implement greywater recycling systems to reuse water from sinks, showers, and washing machines for non-potable purposes like landscape irrigation. Proper treatment ensures the safe and efficient reuse of greywater.

d. Water-Efficient Appliances:

Upgrade to water-efficient appliances and fixtures. High-efficiency toilets, low-flow faucets, and water-saving washing machines contribute to reduced water consumption without compromising functionality.

Permaculture Design Principles:

Permaculture is a holistic approach to designing sustainable and self-sufficient ecosystems. Applying permaculture principles to your homestead enhances its resilience, biodiversity, and overall sustainability. Consider the following permaculture design elements:

a. Polyculture Planting:

Embrace polyculture by planting diverse crops together. Polyculture fosters natural pest control, nutrient cycling, and healthier soil, reducing the need for synthetic fertilizers and pesticides.

b. Companion Planting:

Incorporate companion planting techniques to enhance plant health and yield. Certain plant combinations provide mutual benefits, such as deterring pests or improving soil fertility.

c. Keyhole Gardens and Swales:

Implement keyhole gardens and swales to conserve water and optimize irrigation. These design elements capture and distribute rainwater efficiently, promoting sustainable water use in your garden.

d. Food Forests:

Design and cultivate food forests, mimicking natural ecosystems with layers of edible plants. Food forests provide a sustainable and resilient source of fruits, nuts, herbs, and vegetables.

Waste Reduction and Recycling Practices:

Minimizing waste and implementing recycling practices contribute to a green homestead. Reduce your environmental impact by adopting the following waste reduction strategies:

a. Composting Systems:

Set up composting systems to recycle organic waste from kitchen scraps and garden trimmings. Compost enriches the soil, reduces the need for chemical fertilizers, and minimizes the amount of waste sent to landfills.

b. Recycling Stations:

Establish recycling stations for materials like glass, plastic, paper, and metal. Proper sorting and recycling prevent these materials from ending up in landfills, conserving resources and reducing pollution.

c. Upcycling and Repurposing:

Embrace upcycling and repurposing as creative ways to give new life to old or discarded items. Transforming materials into new, functional items reduces the demand for new resources and minimizes waste.

d. Minimal Packaging and Bulk Buying:

Choose products with minimal packaging to reduce overall waste. Additionally, consider bulk buying to minimize packaging waste and save on costs. Purchase goods in larger quantities to reduce the need for excess packaging.

Natural Building Techniques:

Natural building techniques prioritize the use of locally sourced, sustainable, and non-toxic materials. Incorporate these techniques to create structures that are environmentally friendly and energy-efficient:

a. Earthen Construction:

Explore earthen construction methods such as adobe, cob, or rammed earth. These techniques utilize natural materials, have low environmental impact, and provide excellent insulation properties.

b. Straw Bale Construction:

Consider straw bale construction for energy-efficient and well-insulated walls. Straw bales, when properly sealed, offer effective insulation and contribute to a comfortable indoor environment.

c. Living Roofs and Walls:

Integrate living roofs and walls to enhance insulation and promote biodiversity. Green roofs and walls provide additional insulation, reduce stormwater runoff, and create habitats for plants and wildlife.

d. *Natural Finishes:*

Choose natural finishes for interiors, such as clay plaster, lime wash, or milk paint. These finishes are non-toxic, breathable, and contribute to a healthier indoor environment compared to conventional synthetic options.

Educating and Engaging the Community:

The journey towards a green homestead extends beyond your property boundaries. Educating and engaging the local community fosters a broader understanding of sustainable living and promotes collective action. Consider the following strategies:

a. Workshops and Seminars:

Organize workshops and seminars on sustainable living topics. Share your knowledge and experiences with the community, covering aspects like energy efficiency, permaculture, and eco-friendly building practices.

b. Demonstration Projects:

Implement demonstration projects on your homestead. Showcase sustainable building techniques, renewable energy systems, or permaculture designs. Hands-on demonstrations provide tangible examples for the community.

c. Community Gardens and Shared Resources:

Support the establishment of community gardens and shared resources. Collaborative initiatives like tool-sharing programs, seed libraries, or communal composting encourage sustainable practices among community members.

d. Collaboration with Local Organizations:

Collaborate with local environmental organizations, schools, or community centers. Partnering with these entities amplifies your impact, reaching a broader audience and fostering a sense of collective responsibility.

Transforming your homestead into a green and eco-friendly haven involves a holistic approach that encompasses sustainable building practices, energy efficiency, water conservation, permaculture design, waste reduction, and community engagement. By implementing these simple steps, you not only reduce your environmental footprint but also contribute to a more resilient, self-sufficient, and harmonious way of living on the land. Embrace the principles of a green homestead and cultivate a lifestyle that honors the Earth and inspires positive change within your community.

CHAPTER 9:
HARMONY ON THE HOMESTEAD: CREATING A BALANCED ECOSYSTEM

Achieving harmony on the homestead involves recognizing and fostering the interconnectedness of plants, animals, and the environment. This chapter explores the intricate web of relationships within a homestead ecosystem, emphasizing the importance of balance and sustainability. From understanding the roles of key species to implementing permaculture principles, this comprehensive guide aims to help homesteaders create a thriving and balanced ecosystem that nurtures both the land and its inhabitants.

Understanding the Interconnectedness of Plants, Animals, and Environment

Ecological Principles on the Homestead:

To create a balanced ecosystem, it's essential to grasp foundational ecological principles. These principles guide the relationships between living organisms and their environment, forming the basis for sustainable homesteading practices:

a. Biodiversity:

Encourage biodiversity by cultivating a variety of plants and providing habitats for diverse animal species. A diverse ecosystem is more resilient to pests, diseases, and environmental changes.

b. Interdependence:

Recognize the interdependence of species within the ecosystem. Plants, animals, and microorganisms often rely on each other for resources, such as pollination, nutrient cycling, and pest control.

c. Succession:

Understand ecological succession, the natural process of changes in plant and animal communities over time. Design your homestead with an awareness of succession, allowing for the dynamic evolution of the ecosystem.

d. Regenerative Practices:

Adopt regenerative practices that enhance soil health, conserve water, and promote the overall well-being of the ecosystem. Techniques like cover cropping, rotational grazing, and agroforestry contribute to regenerative agriculture.

Roles of Key Species in the Ecosystem:

Every species on the homestead plays a specific role in maintaining balance. Understanding the roles of key species helps you design a holistic and resilient ecosystem:

a. Pollinators:

Recognize the importance of pollinators, such as bees, butterflies, and birds. These creatures facilitate the reproduction of flowering plants, ensuring a bountiful harvest of fruits and vegetables.

b. Predators and Pest Control:

Welcome natural predators like ladybugs, spiders, and predatory insects. These beneficial organisms help control pest populations, reducing the need for chemical interventions.

c. Decomposers:

Appreciate the role of decomposers, including earthworms, fungi, and bacteria. Decomposers break down organic matter, returning nutrients to the soil and supporting overall soil health.

d. Nitrogen Fixers:

Integrate nitrogen-fixing plants like legumes into your garden. These plants form symbiotic relationships with nitrogen-fixing bacteria, enhancing soil fertility by converting atmospheric nitrogen into a form usable by plants.

Permaculture Design for Ecosystem Harmony:

Permaculture principles provide a holistic framework for designing harmonious ecosystems. Implementing permaculture on the homestead fosters sustainability, resilience, and a thriving balance among plants, animals, and the environment:

a. Zoning:

Apply zoning principles to organize your homestead based on the frequency of human interaction and the needs of different elements. For example, place frequently used and intensively managed areas closer to the home for easy access.

b. Guild Planting:

Create plant guilds, or companion planting arrangements, to maximize beneficial interactions between plants. Companion plants can provide support, shade, or pest resistance to each other.

c. Keyhole Design:

Implement keyhole designs in garden layouts to minimize paths and maximize growing space. Keyhole gardens reduce the need for excessive walkways, promoting efficient use of space and resources.

d. Water Harvesting:

Harvest rainwater and design water-capturing features like swales to retain and distribute water effectively. Efficient water use is integral to permaculture design, fostering a sustainable approach to irrigation.

Regenerative Agriculture Practices:

Regenerative agriculture aims to restore and enhance the health of the land. Implementing regenerative practices contributes to ecosystem harmony on the homestead:

a. Holistic Grazing:

Practice holistic grazing by rotating livestock through different paddocks. This approach prevents overgrazing, allows for natural regeneration of pastures, and promotes soil health.

b. Cover Cropping:

Utilize cover crops during periods of fallow or between main crops. Cover crops protect and improve the soil, prevent erosion, and enhance biodiversity.

c. No-Till Gardening:

Adopt no-till gardening methods to preserve soil structure and minimize disruption to the ecosystem. No-till practices retain beneficial microorganisms and enhance water retention.

d. Comprehensive Crop Planning:

Plan your crop rotations strategically to avoid depleting soil nutrients. Rotate crops to break pest and disease cycles, improve soil fertility, and support overall ecosystem health.

Forest Gardening and Agroforestry:

Embracing forest gardening and agroforestry principles brings the benefits of a forest ecosystem to your homestead. These practices integrate trees, shrubs, and crops, mimicking natural forest dynamics:

a. Food Forest Design:

Design food forests with layers of trees, shrubs, and ground-cover plants. Food forests mimic natural ecosystems, providing a diverse array of edible plants and supporting wildlife.

b. Agroforestry Systems:

Incorporate agroforestry systems, such as alley cropping or silvopasture, to diversify your homestead. These systems integrate trees and other crops or livestock, promoting mutual benefits and sustainable land use.

c. Perennial Polycultures:

Integrate perennial polycultures, combining perennial plants in diverse combinations. Perennials contribute to long-term soil health, reduce the need for annual replanting, and support ecosystem stability.

d. Wildlife Habitat Enhancement:

Enhance wildlife habitat within your agroforestry or forest gardening systems. Native trees and shrubs attract a variety of wildlife, creating a balanced ecosystem that benefits both plants and animals.

Wildlife Conservation and Habitat Enhancement:

Actively participating in wildlife conservation and enhancing habitat on your homestead contributes to biodiversity and ecosystem health:

a. Native Plantings:

Prioritize native plantings to create habitats for local wildlife. Native plants provide food and shelter for insects, birds, and other fauna, fostering a resilient and balanced ecosystem.

b. Pond and Wetland Creation:

Develop ponds or wetlands to attract amphibians, insects, and waterfowl. These features enhance biodiversity, support pollinators, and provide habitat for various aquatic species.

c. Hedgerows and Wildlife Corridors:

Establish hedgerows and wildlife corridors to connect different areas of your homestead. These features enable the movement of wildlife, contribute to genetic diversity, and enhance ecosystem resilience.

d. Nesting Boxes and Habitat Features:

Install nesting boxes, bat houses, and other habitat features to support specific wildlife species. Creating designated spaces for nesting and shelter encourages the presence of beneficial animals on your homestead.

Seasonal and Crop Rotation Strategies:

Adapting your homestead practices to the changing seasons and implementing crop rotation strategies contributes to a balanced and sustainable ecosystem:

a. Seasonal Planning:

Plan your homestead activities and plantings according to seasonal cycles. Consider factors such as temperature, sunlight, and precipitation to optimize planting and harvesting schedules.

b. Crop Rotation Benefits:

Practice crop rotation to prevent soil-borne diseases, manage pests, and improve soil fertility. Rotate crops strategically to break pest and disease cycles while maintaining a healthy balance in the ecosystem.

c. Intercropping and Succession Planting:

Embrace intercropping and succession planting to maximize space and resources. These practices involve planting different crops in proximity or successively, promoting biodiversity and efficient use of growing areas.

d. Fall and Winter Cover Crops:

Plant cover crops during fall and winter to protect and enrich the soil. Cover crops prevent erosion, add organic matter to the soil, and contribute to the overall health of the ecosystem during periods of rest.

Holistic Animal Husbandry Practices:

Balancing the needs of animals with the health of the ecosystem requires adopting holistic animal husbandry practices:

a. Rotational Grazing:

Implement rotational grazing systems to prevent overgrazing and promote natural forage regeneration. Well-managed rotational grazing benefits both livestock and the pasture ecosystem.

b. Animal Integration in Agroecosystems:

Integrate animals into agroecosystems to mimic natural ecological processes. For example, chickens can be integrated into orchards to control pests, providing a symbiotic relationship.

c. Composting Livestock Manure:

Properly compost livestock manure to create nutrient-rich compost for the garden. Composting minimizes odors, reduces runoff, and transforms waste into a valuable resource for the homestead.

d. Heritage Breeds and Biodiversity:

Consider raising heritage breeds of livestock to contribute to genetic diversity. Heritage breeds often have unique traits that make them well-suited to specific environments, enhancing overall ecosystem resilience.

Creating harmony on the homestead involves a thoughtful and holistic approach to ecosystem management. Understanding the interconnectedness of plants, animals, and the environment is the foundation for cultivating a balanced and sustainable homestead. Whether through permaculture design, regenerative agriculture, wildlife conservation, or holistic animal husbandry, each practice contributes to a resilient and thriving ecosystem. By embracing these principles, homesteaders can foster a landscape where every element plays a vital role, contributing to the health and vitality of the entire homestead ecosystem.

CHAPTER 10:
HOMESTEADING THROUGH THE SEASONS: A YEAR-ROUND GUIDE

Homesteading is a dynamic journey that unfolds throughout the seasons, each presenting its own challenges and opportunities. This chapter explores seasonal strategies for maximized productivity on the homestead, guiding you through year-round activities that align with the ebb and flow of nature. From planning your garden in spring to winterizing your homestead in preparation for colder months, this comprehensive guide aims to help homesteaders make the most of every season for a thriving and sustainable lifestyle.

Seasonal Strategies for Maximized Productivity

Spring Awakening: Planning and Planting:

Spring is a season of renewal and growth, making it a crucial time for homesteaders to lay the groundwork for the rest of the year. Consider the following strategies:

a. Garden Planning:

Start the season by planning your garden layout. Consider crop rotation, companion planting, and the integration of permaculture principles to optimize the use of space and resources.

b. Seed Starting:

Begin indoor seed starting for crops that require a longer growing season. This ensures robust and healthy seedlings ready for transplanting once the risk of frost has passed.

c. Soil Preparation:

Prepare garden beds by amending the soil with compost and organic matter. Spring is an ideal time to test soil pH, address nutrient deficiencies, and create optimal growing conditions.

d. Fruit Tree Care:

Attend to fruit trees by pruning, fertilizing, and inspecting for any signs of disease or pests. Proper care in spring sets the stage for a bountiful harvest later in the year.

Summer Bounty: Cultivating and Harvesting:

Summer is the season of abundance, where the hard work put into planning and planting during spring comes to fruition. Make the most of this productive period with the following strategies:

a. Intensive Gardening:

Practice intensive gardening techniques to maximize yield. Interplanting, vertical gardening, and succession planting can help you make the most of available space.

b. Harvesting and Preservation:

Begin harvesting crops as they reach maturity. Implement preservation methods such as canning, freezing, and drying to store surplus produce for the upcoming seasons.

c. Livestock Management:

Monitor and manage livestock to ensure their well-being during the warmer months. Provide ample shade, fresh water, and adjust feeding schedules based on increased forage availability.

d. Pest Control:

Stay vigilant for pests that may affect your crops during the summer. Implement natural pest control methods such as companion planting, beneficial insect release, and handpicking.

Fall Transition: Planning and Harvest Continuation:

As summer transitions into fall, it's time to plan for the coming seasons and extend the harvest. Consider the following strategies:

a. Fall Garden Planting:

Plant cool-season crops for a fall harvest. Vegetables like kale, carrots, and Brussels sprouts thrive in the cooler temperatures of autumn.

b. Cover Cropping:

Utilize cover crops on fallow areas to protect and enrich the soil during the offseason. Cover crops help prevent erosion and add organic matter to the soil.

c. Preserving the Harvest:

Continue preserving the late summer and fall harvest. Experiment with new preservation methods like fermenting or pickling to diversify your pantry.

d. Preparing for Frost:

Monitor weather forecasts and be prepared for the first frosts. Cover sensitive plants, harvest remaining warm-season crops, and store or process them to avoid losses.

Winter Planning: Reflecting and Strategizing:

Winter provides a valuable opportunity for reflection, planning, and strategic preparation for the coming seasons. Use this time wisely with the following strategies:

a. Homestead Review:

Reflect on the successes and challenges of the past year. Use this insight to adjust your homesteading plans, crop selections, and overall strategy for the upcoming seasons.

b. Crop Rotation Planning:

Plan crop rotations for the next year to prevent soil-borne diseases and maintain soil fertility. Consider the specific needs and interactions of different crops within your rotation.

c. Infrastructure Maintenance:

Use the winter months for essential maintenance tasks. Inspect and repair infrastructure, such as fences, outbuildings, and equipment. This proactive approach ensures everything is in good working order when spring arrives.

d. Education and Skill Development:

Invest time in learning new skills or deepening existing ones. Winter is an ideal period for reading, attending workshops, and honing your homesteading knowledge and techniques.

Year-Round Livestock Care:

Livestock care is a constant responsibility that requires attention to their unique needs throughout the seasons. Implement these strategies for year-round livestock management:

a. Winter Shelter and Feeding:

Provide adequate shelter, warmth, and a balanced diet for livestock during the colder months. Ensure access to clean water and consider supplemental feeding to meet nutritional requirements.

b. Spring Grazing Rotation:

Plan rotational grazing systems for spring and summer. This ensures that pastures are grazed evenly, preventing overgrazing and promoting regrowth.

c. Summer Heat Management:

Implement strategies to help livestock cope with summer heat. Provide shade, ample water, and consider adjusting feeding schedules to avoid the hottest parts of the day.

d. Fall and Winter Health Checks:

Conduct health checks on livestock as fall transitions into winter. Address any issues promptly, and consult with a veterinarian for vaccinations and preventative measures.

Greenhouse Gardening and Winter Harvests:

Extend your growing season and enjoy fresh produce even during the colder months with greenhouse gardening and strategic winter harvests:

a. Greenhouse Maintenance:

Maintain and prepare your greenhouse for winter use. Clean and repair structures, check heating systems, and insulate as needed to create a conducive environment for winter gardening.

b. Winter Crop Selection:

Choose cold-hardy crops for winter greenhouse cultivation. Leafy greens, root vegetables, and herbs can thrive in the protected environment of a greenhouse during colder months.

c. Winter Harvesting Techniques:

Implement techniques such as succession planting, protective row covers, and cold frames to extend outdoor harvesting into the winter months. Harvesting root vegetables before the ground freezes ensures a fresh supply.

d. Indoor Herb Gardens:

Cultivate indoor herb gardens on sunny windowsills or under grow lights. Fresh herbs add flavor to winter dishes and can be easily grown indoors with proper care.

Seasonal Beekeeping Practices:

Beekeeping requires seasonal adjustments and specific practices to ensure the health and productivity of your bee colonies. Implement these strategies for year-round beekeeping:

a. Spring Hive Inspections:

Conduct thorough hive inspections in spring to assess colony strength, queen health, and resource availability. Address any issues and provide supplemental feeding if necessary.

b. Summer Honey Harvest:

Harvest honey during the summer months when nectar flow is at its peak. Use proper beekeeping techniques to extract honey while leaving enough for the bees to sustain themselves.

c. Fall Hive Preparation:

Prepare hives for winter by ensuring they have sufficient honey stores and protecting against cold drafts. Consider insulating hives in colder climates to help bees survive winter.

d. Winter Hive Monitoring:

Monitor beehives during winter to ensure they have enough food reserves. In colder regions, consider using hive wraps or other insulation methods to protect against extreme cold.

Community Engagement and Collaboration:

Engaging with the local community and collaborating with fellow homesteaders enhances the homesteading experience and creates a network of support. Consider these strategies for community involvement:

a. Seasonal Workshops and Events:

Host or participate in seasonal workshops and events. Share your knowledge and learn from others, creating a supportive community of like-minded individuals.

b. Seasonal Farm-to-Table Experiences:

Offer farm-to-table experiences during the growing season. Open your homestead to the community for activities like pick-your-own events, farm tours, or seasonal markets.

c. Community Seed Swaps:

Participate in or organize community seed swaps. Exchanging seeds with neighbors promotes biodiversity and ensures a varied selection of crops for everyone involved.

d. Collaborative Projects:

Collaborate with other homesteaders on larger projects. This could include bulk buying, shared infrastructure like communal tool sheds, or joint efforts in areas like composting or animal husbandry.

Homesteading through the seasons involves a dynamic and strategic approach that aligns with the natural rhythms of the year. By embracing seasonal strategies for planning, planting, cultivating, and reflecting, homesteaders can maximize productivity, ensure sustainability, and create a balanced and fulfilling life on the land. Whether tending to gardens, caring for livestock, or engaging with the community, each season offers unique opportunities for growth, learning, and the joy of a well-nurtured homestead.

CHAPTER 11:
BALANCING ACT: WORK, LIFE, AND SUSTAINABLE HOMESTEADING

Sustainable homesteading is a rewarding yet demanding lifestyle that requires a delicate balance between work and personal life. Achieving harmony while nurturing a thriving homestead involves strategic planning, time management, and a deep understanding of one's priorities. In this chapter, we explore the challenges and strategies for finding equilibrium in the midst of a bustling homestead, ensuring that the pursuit of sustainability enhances, rather than hinders, the overall quality of life.

The Homesteader's Journey: Navigating Challenges and Rewards

Embarking on a homesteading journey is a courageous and fulfilling endeavor. The rewards of producing your own food, living closer to nature, and fostering sustainability are substantial. However, the journey also presents challenges that can impact the delicate balance between work, life, and homesteading responsibilities.

Defining Your Homesteading Goals:

Before delving into the daily demands of homesteading, it's crucial to define your goals clearly. What aspects of homesteading are most important to you? Whether it's achieving self-sufficiency, minimizing environmental impact, or enjoying a simpler way of life, establishing priorities will guide your decisions and help maintain balance.

a. Self-Reflection:

Take time for self-reflection to identify your values and aspirations. Understanding what motivates your homesteading journey will inform the goals you set for your sustainable lifestyle.

b. Long-Term Vision:

Develop a long-term vision for your homestead. This vision will act as a compass, guiding your decisions and actions toward the sustainable and balanced lifestyle you aim to achieve.

c. Flexibility and Adaptability:

Be open to adjusting your goals as your homesteading journey unfolds. Flexibility and adaptability are essential as you learn more about your land, resources, and personal preferences.

d. Balancing Aspirations:

Strike a balance between ambitious aspirations and practical considerations. While it's natural to have grand visions, be realistic about the time and effort required to achieve them while maintaining a harmonious lifestyle.

Time Management Strategies:

Time is a valuable resource on the homestead, and effective time management is crucial for maintaining a balance between work, life, and sustainable practices.

a. Prioritizing Tasks:

Prioritize tasks based on urgency and importance. Recognize the seasonal nature of homesteading activities and allocate time accordingly, giving priority to essential tasks during critical periods.

b. Time Blocking:

Implement time-blocking techniques to allocate specific periods for different activities. Designate time for gardening, livestock care, and personal pursuits, ensuring a well-rounded schedule.

c. Setting Realistic Expectations:

Establish realistic expectations for your daily and weekly workload. Avoid overcommitting, as this can lead to burnout and compromise the overall enjoyment of homesteading.

d. Delegate and Share Responsibilities:

If you have a family or community members on your homestead, delegate tasks and share responsibilities. Collaborative efforts not only lighten the workload but also foster a sense of shared purpose.

Creating Functional Homestead Zones:

Organizing your homestead into functional zones streamlines activities and contributes to a more balanced lifestyle.

a. Home Zone:

Designate areas around your home for relaxing and recreational activities. Create spaces for family time, hobbies, and leisure to ensure a healthy work-life balance.

b. Productive Zone:

Organize productive zones for gardening, livestock, and other homesteading activities. Plan these areas efficiently, considering proximity to water sources, sunlight, and ease of access.

c. Workshop and Storage Zone:

Establish dedicated spaces for workshops, tool storage, and equipment maintenance. Keeping these areas organized reduces time spent searching for tools and enhances overall efficiency.

d. Restorative Zone:

Incorporate restorative zones for contemplation and connection with nature. These areas can serve as retreats for relaxation and reflection, contributing to mental well-being.

Integration of Technology and Modern Tools:

Embracing technology and modern tools can enhance efficiency on the homestead, freeing up time for personal pursuits.

a. Smart Gardening Tools:

Explore smart gardening tools that automate watering schedules, monitor soil conditions, and optimize plant care. These tools save time and contribute to more efficient gardening practices.

b. Livestock Monitoring Systems:

Implement livestock monitoring systems that provide real-time data on the health and well-being of animals. Automated feeding systems and smart fencing contribute to streamlined livestock management.

c. Energy-Efficient Appliances:

Invest in energy-efficient appliances for tasks like food preservation and cooking. These appliances reduce the time spent on daily chores while minimizing environmental impact.

d. Homesteading Apps:

Utilize homesteading apps for planning, record-keeping, and resource management. These digital tools can help streamline tasks, providing a more organized and efficient homesteading experience.

Cultivating a Balanced Homesteading Lifestyle

Achieving a balanced homesteading lifestyle goes beyond managing tasks; it involves cultivating a mindset that prioritizes well-being, sustainability, and the enjoyment of the homesteading journey.

Mindful Homesteading Practices:

Infuse mindfulness into your homesteading practices to enhance your connection with the land and foster a more balanced lifestyle.

a. Seasonal Awareness:

Cultivate awareness of seasonal changes and their impact on your homestead. Align your activities with the natural rhythms of the land, allowing for a more harmonious relationship with your environment.

b. Observation and Reflection:

Practice regular observation and reflection on your homestead. This mindfulness allows you to notice subtle changes, address issues promptly, and appreciate the evolving beauty of your surroundings.

c. Mindful Animal Care:

Extend mindfulness to the care of your animals. Pay attention to their behavior, health, and well-being, fostering a deeper connection and understanding between you and your livestock.

d. Gratitude Practices:

Cultivate gratitude for the abundance your homestead provides. Take moments to appreciate the fruits of your labor, the beauty of nature, and the sense of self-sufficiency that homesteading brings.

Family and Community Involvement:

Homesteading can be a family or community affair, fostering a sense of shared responsibility and enriching relationships.

a. Family Meetings:

Hold regular family meetings to discuss homesteading goals, responsibilities, and individual needs. This collaborative approach ensures that everyone feels engaged and valued in the homesteading journey.

b. Involving Children:

Involve children in age-appropriate homesteading activities. This not only teaches valuable skills but also instills a sense of responsibility and a connection to the land from an early age.

c. Community Collaboration:

Collaborate with neighbors or fellow homesteaders for shared projects or resources. Community involvement creates a supportive network and provides opportunities for learning and skill-sharing.

d. Balancing Personal and Shared Spaces:

Define personal and shared spaces within your homestead. Balancing private retreats with communal areas ensures that each family member has a balance between personal space and shared responsibilities.

Holistic Wellness on the Homestead:

Prioritize holistic wellness to ensure that homesteading contributes to physical, mental, and emotional well-being.

a. Physical Health Practices:

Incorporate physical health practices into your routine, such as regular exercise, stretching, and ergonomic considerations during homesteading activities. This contributes to overall well-being and prevents injuries.

b. Mental Health Awareness:

Cultivate awareness of mental health and stress levels. Practice stress-reducing activities such as meditation, nature walks, or creative pursuits to maintain a healthy mindset.

c. Rest and Recovery:

Prioritize sufficient rest and recovery. Adequate sleep breaks during the day, and scheduled downtime contribute to sustained energy levels and prevent burnout.

d. Emotional Resilience:

Develop emotional resilience to navigate the inevitable challenges of homesteading. Cultivate a positive mindset, seek support when needed, and acknowledge that setbacks are a natural part of the homesteading journey.

Continuous Learning and Adaptation:

Homesteading is a dynamic and ever-evolving lifestyle. Embrace a mindset of continuous learning and adaptation to stay resilient and maintain a healthy balance.

a. Learning from Challenges:

View challenges as opportunities for growth and learning. Reflect on setbacks, assess what can be improved, and apply lessons learned to enhance your homesteading practices.

b. Skill Development:

Dedicate time to acquiring new skills that align with your homesteading goals. Continuous skill development not only enriches your abilities but also adds variety to your daily tasks.

c. Adapting to Change:

Be open to adapting your homesteading practices based on changing circumstances. Whether it's shifts in weather patterns, market demand, or personal priorities, flexibility is key to maintaining balance.

d. Community Learning Experiences:

Participate in community learning experiences, such as workshops, classes, or farm tours. Engaging with a broader community of homesteaders provides fresh perspectives and a supportive network for ongoing learning.

Conclusion: Nurturing a Thriving Homestead and Life Balance

Balancing work, life, and sustainable homesteading is an ongoing process that requires intention, mindfulness, and a willingness to adapt. By defining clear goals, implementing effective time management strategies, cultivating a balanced mindset, and prioritizing holistic wellness, homesteaders can enjoy the rewards of sustainable living without sacrificing the quality of life. The key lies in creating a harmonious synergy between the demands of the land and the fulfillment of personal and family needs, ultimately nurturing a thriving homestead and life balance.

www.ingramcontent.com/pod-product-compliance
Lightning Source LLC
Chambersburg PA
CBHW071052290526
45795CB00004B/1449